Why round up the wild horses?

This question is asked of us many times; it's probably the first question asked by individuals that read about what we are doing. Our work is based on the belief that throughout the wild bands of horses roaming the Western United States of America there are some that still carry the high percentage of DNA from the horses brought to the New World by the Spanish Explorers. We know that most of these animals are not found roaming with the typical bands on government lands. The horses that we are trying to find fall into the very rare range, maybe 1 in 30, and are located in isolated areas of the West where access is difficult, dangerous, and almost impossible.

1. These animals are the only remaining link to the original horses that came back to the Americas after a 6000-year hiatus.
2. These animals changed the Native Americans way of life and redefined their culture.
3. These are the animals on whose back this nation was built.

History tells us that without the wild horse, the face of America would not be what it is today

Imagine,
- a horseless Native American,
- the Pony Express on foot,
- the first cowboys walking behind their cattle
- the West without the Mustang.

We know why we want to preserve these magnificent creatures; now you must ask yourself:

Why do you want to protect them?

The most important aspect of a round up of wild horses or a free-range horse round up is the first constant:

Wild Horses want to remain free.

There are two dynamics of this constant:

1: The animals desire not to be controlled.

- The animals will remain true to the first constant, and based on their instinctive reactions, "the fight or flight syndrome", their actions are predictable.
- In most instances, the reaction will be that of flight, but in rare instances, the fight reactions may come into play.
- An attempt to escape confinement by running through or by its capturers is not to be considered the fight reaction, but the flight reaction to confinement.

2: The wranglers desire to control the animals.

- This dynamic, at its best, is unpredictable, almost to the person. Each round-up will be different. This may be because of the humans' reaction to the fight or flight syndrome.
- We humans also adhere to the syndrome but in most instances react very differently, especially when it comes to controlling animals that we consider pets.
- One must remember that because of the nature of the relationship between humans and equines, the role of humans and the wild horse becomes blurred. As humans, we look at the horse not with the eyes of a predator but with the eyes of a companion. The wild horse, on the other hand, looks at humans with the eyes of a prey animal, and in their view, we are predators.

Note: Wild horses are to domestic horses as wolves are to dogs.

- We can never forget that these creatures are wild, and our relationship with them must be cautious and calm. We tend to forget this fact and treat the wild horses in the same manner that we would treat a domestic horse.
- This is why we have established our own method of dealing with wild horses and our own method of rounding up these horses. We do not claim to have all the answers and in some instances, because of circumstances, we may modify these methods, but we try to adhere to the following procedures as closely as possible.

.

How to tell the condition and health of a band of wild horses.

Most individuals seeing wild horses for the first time consider them to be in poor condition. This may be the case, especially for a band of wild horses near a roadway or pathway. When wild horses stay close to areas occupied by humans this suggests they are experiencing trauma caused by environmental conditions, such as drought, famine, or human interference.

Wild horses will generally only allow humans to approach within a quarter of a mile. Sometimes a human on horseback may be able to get closer, but again a limit to the nearness is established. Once this line is crossed, the horses will move to what they consider a safe distance. It will take days and sometimes weeks for an individual to get near enough to a band of wild horses to tell the condition of the band. However, there are ways of telling the general condition of the band without disturbing their daily routine.

1. Check the band's stools or droppings. If the droppings are firm and round, this is a good sign. If the droppings are runny or more like cow droppings, this may indicate parasitic infection.
2. Look at the hoof prints of the band; generally, the hooves will appear in good shape with very few nicks or chips. If the prints show long growth or flaring, this is not a good sign.
3. If hair from the band can be found on brush, bushes, or on the ground, check the texture and pliability. Good texture and good pliability usually indicates a healthy band.

When you are able to get close enough to the band to be able to distinguish the individuals' ears you can get a better idea of the band's condition. In most instances, you will see that the animals are what is called "ribby." This is not cause for alarm; most equines that survive in the wild do not overeat and are lean, just like an athlete, a runner, or a swimmer. The only time this condition is of any concern is when it is combined with sunken hip muscles, where you can see the points of the hips. Be careful when using this criterion especially if you are dealing with an older horse. Another obvious give away to poor condition is the state of the mane and tail. Very rarely will you see a healthy wild horse with a ratted mane or tail. I do not know why this is, but only the horses that are sick or in poor condition have ratted manes or tails. Even on free-range horses, this is something that I look at seriously. When looking at the animal's coat, it looks unhealthy if it has no sheen or is dull and lifeless. The general thought is that a horse with a dull, lifeless, coat needs to be wormed. This usually is not the case in wild horses, it's not that they don't get worms, it's just that it is a temporary condition and gone within days and at the most two weeks. Wild and free-range horses do not eat where they have defecated, unlike domestic horses. .

The care and maintenance of the wranglers.

When the horses are rounded up, they need to be cared for and kept under the least stressful conditions possible. The wranglers also must be maintained under the least stressful conditions possible. Usually, when we tell people what we do, the reply is, "That sounds like a great job." You can hear the envy in their voices and see it in their mannerisms. What they don't realize is the work that goes into becoming a wrangler. Usually, they only think of the romantic side of the adventure. They see the mixing of the wild mystic and the old days of the west. The wrangler is the door to the old ways when most believe things were simpler. Since times have changed, some of the simple things are not that simple anymore.
Listed below are some are items needed for the care and maintenance of a wrangler.

- The head must be covered, of course, most wranglers will wear their western hat, and if this is done at least wear a wind strap so that if the hat comes off the horses are not spooked. Any type of head cover will do, just make sure the head is covered.
- The body is next; make sure you dress in layers because it is much easier and quicker to take clothes off than to put them on. When you are on a spotting expedition, the color of clothes may become important. Earth tones are always best when working with wilds.
- The feet are the most important aspect of the wrangler; especially when all the work is done on the ground. Comfortable walking shoes or boots is a must. If you wear walking shoes, then you must wear gators (Protectors that wrap around one's ankles and calf muscles to prevent snake bites.).
- Now that we have taken care of the outside of the body remember that the inside needs to be also protected, bring plenty of drinking water and maybe some protein bars along with you.
- The last thing about your body: **do not wear any perfumes or aftershave**. If anything, take some of the local flora, maybe sage depending where you are, and rub it on your clothes. This will help to mask your odor. Wild horses will smell you long before you see them, or they see you.

Probably the most important thing to remember is to know your physical limits and respect them. Do not try to maintain pace with the horses. You should stop every 20 to 30 minutes and rest for at least 5 minutes or longer during the summer months.

Our wild horse round-ups are divided into three phases. No one phase is more important than the other.

- **The first, is the spotting and counting phase**: This begins with locating the individual bands.

- **The second is the physical round up phase**. This phase takes the most endurance because it done on foot. We "walk the horses down."

- **The third and final is the taking of blood (DNA) phase**. For us Dr. Gus Cothran is provided the blood samples and analyzes them at his laboratory.

Phase One

A: Spotting and counting

The importance of this part of the operation is to be able to see physically the animals that are going to be rounded up and to lay out the strategy for the roundup. Sometimes we can tell by looking at the horses if they will qualify for our project. We look for horses that do not exceed 15 hands in height. We have not, to date, found any horses over 15 hands high that qualify for us*. If we find this feature to be predominant throughout the herd, we do not pursue the round up any further. But if the general height of the band is under 15 hands in height we go on to the next step, photographing and videotaping each band and each horse within the band. This gives us an idea of how many horses are actually in the herd and how many bands they are divided into throughout the herd. It is important not to disturb the horses at this point of the operation if the horses are constantly moving away from it is hard to obtain good pictures. That is why it is a necessity to have a camera with a telephoto lens, at least a 500 or a 1000 mm. lens. The video camera should be set up with a telephoto lens also.

During this part of the operation, along with getting pictures, you are gaining the trust of the band, not to any great extent but at least to a point of toleration. This is important because the progress of each phase depends on the success of the previous one. If the horses do not allow you to share some of their space, you will never be in a position to have a non-violent roundup. If done this assures no one neither human nor animal gets hurt. It is also imperative that the horses never consider you a threat. If you are considered a threat to the band, you might as well pack it in and go on to the next band. You have, at least for the time being, lost any hope of rounding that band up without someone getting hurt. Once you have taken and developed your pictures and know the different bands you can go on to the next step that we call the "push."

B: The Push

You were unknowingly doing this when you were trying to get your pictures. Did you notice that when you tried to get close enough to get a good picture you were only allowed to get so close; then the horses would move. This was the start of what is called the "push" or "pushing." When a push is performed, the animals will move the same distance away from the pusher as the pusher has crossed the **line of demarcation.** You may have noticed that once you stopped moving toward the horses, they stopped moving away. You now should have a general idea of how close you can get to each band without disturbing their routine. **Write it down because this is important**. List each band and the distance from each band you are allowed without crossing the line of demarcation. By now your notebook should have each band (name optional), each lead stallion, and each lead mare listed and the line of demarcation for each band. With this information, you are ready to start pushing. The reason for the push is to see where the horses go when on the move.

*We are looking for horses that have a stronger chance of being related to the original Spanish horse brought to the New World in the 1500's.

The questions you must answer are:

A: What direction do the horses move?

B: What trails do they take when moving?

C: How fast do they move on that trail?

The First Push.

It is important to know what direction the horses move when being pushed, so you can get an idea of what they are doing. Are they just trying to get away and moving randomly, or are they trying to find a hiding place out of view of the pushers? If they just move randomly, go on to the next band and come back to this band later. If they are trying to hide or elude the pushers, mark the trail they use. It would be wise at this point to only push this band a short distance and back off. The distance is important because if you get too close to the band, you are no longer the pusher but an unwanted member of the band and you have defeated the purpose of the push and conceded any chance of manipulating this band.
Do not pursue the push very far the first time. The first push should be calm, short push, nothing more. Do this with every band and log the results.

Repeat this push two or three times.

Log report for Suwannee Band (First Push)

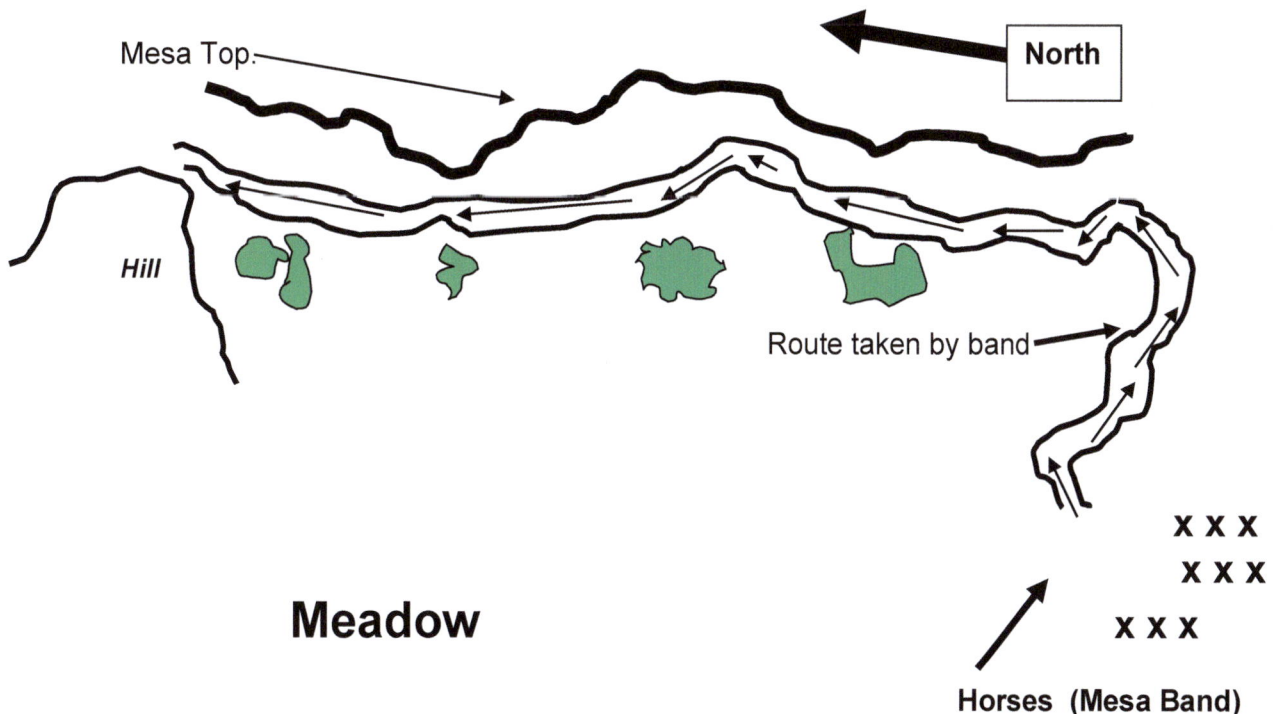

Mesa Top

North

Hill

Route taken by band

Meadow

X X X
X X X
X X X

Horses (Mesa Band)

We photographed the band moving north along a semi tree/bush lined path between an escarpment and a large valley. The horses were in a meadow near the edge of the valley when we moved in to get our pictures. We got to around the 500-foot mark when the black horse with the white patch on its side stopped grazing and looked at us.

We could see that a bay horse was always watching us from the time we got out of our trucks. We took a few more steps, about 10 feet, and the black horse moved away causing the other horses to follow. They only moved what appeared to be 10 to 20 feet away then stopped and went back to grazing. I moved to my right, which would be south, and Cindy and Paul moved to their left or to the north to see if we could get better pictures of the band. The horses did not move. When I tried to get closer, about 50 feet, all the horses in the band looked up, and the black horse took off at a trot. The other horses followed, and the bay horse brought up the rear, always watching where we were. The band followed an old trail up out of the meadow along the side of a Mesa escarpment. Shrubs and trees hide the trail at various places. It then goes around a hill out of sight. We stopped at this point and went on to another band. We pushed this band too hard.

You can see from these notes that even though we are doing a push, we still are taking pictures. It is always good to have a visual record along with a written record of everything you do.

The Second Push.

This push, like the first push, is to learn the movement and direction of the band when being pursued, but unlike the first push, this push will be for three times the distance of the first one. We found out from the first push that this band would allow us to get within 500 feet before they would move away and at about the 450-foot mark they left the grazing area altogether. We also know from the first push that when moved out of the meadow it will take at least two days for the horses to return. Now that we know the trail that this band uses and have seen where it goes, we must place someone out of sight at the place where they disappeared around the hill. This individual must always stay hidden until the horses pass and are far enough away not to be seen, but close enough to observe the direction traveled. When everyone is in place, we start the push. Again we will write down the route traveled and the way the horses moved.

Questions to answer in this push.

Did they walk away from the pusher and continue walking, or did they trot away?

If the band trotted, when did they start to trot, or did they run away, and when did they start their running?

How close was the pusher when they started to trot or run?

What did the pusher do that caused the band to run?
- Do you know?
- Can you correct it?

If you are not aware of what you did, and it was not an outside force like birds or planes, etc., that caused the band's actions, then you must retrace your movements to try to answer the question.

It is important that you do your best to control your actions so that you don't alarm the band. There are enough external forces that you cannot control that will come into play that will cause unexpected reactions from the band.

If they trotted or ran did they change their path or stay on the same path?

Repeat this push at least twice.

Log report for Suwannee Band (Second Push)

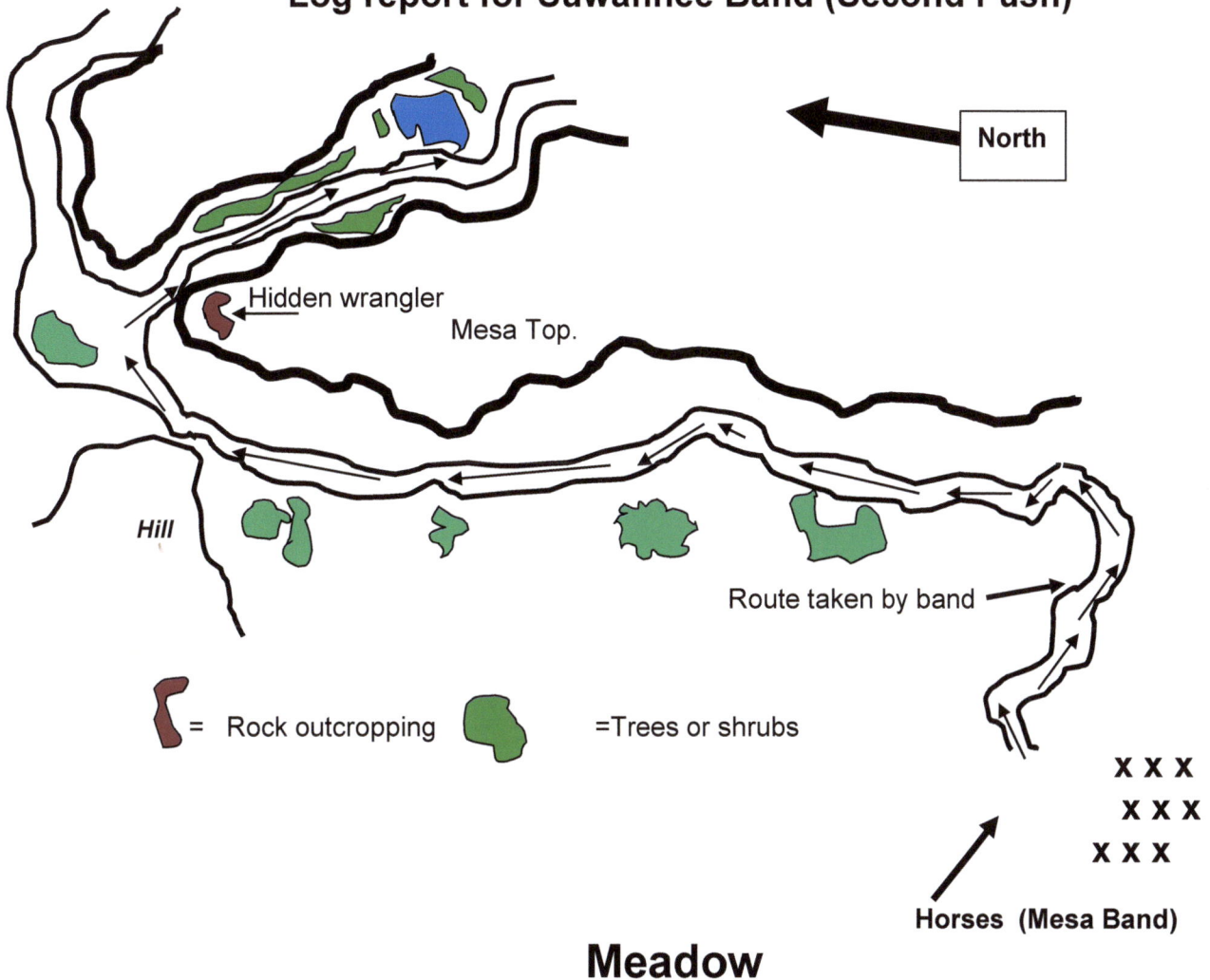

North

Hidden wrangler

Mesa Top.

Hill

Route taken by band

 = Rock outcropping

=Trees or shrubs

X X X

X X X

X X X

Horses (Mesa Band)

Meadow

The Third Push.

This push is the most aggressive push of all and probably the fastest. The idea of this one is to see where the horses go when startled or alarmed.

What route do they feel is the most secure for escaping from predators? Wild horses don't use the same trail for escape when being pushed calmly without fear. However, when threatened, they do have certain escape routes or trails they use that they seem to feel are safe when needed for a quick escape. One important question about this escape route is, **"Does it cross into any other band's territory?"**

This can be noted by looking for **stallion piles.**

> **The way to tell which pile belongs to whom is by watching the mares of each band. They will go to a certain pile and smell it. If the pile is their stallion's they will urinate on it. If it is not their stallions', they will move away from it.**

The importance of this is if the escape route crosses into another band's territory, then it could cause a fight and defeat any chance of a calm round up. The band being pushed becomes the intruder band and may not go back to its territory because the pushers are there. If the band being pushed has started to move into another band's territory, stop the push and go on to another band.

The result of continuing this type of push may cause the band being pushed not to return to its territory for many days or possibly weeks, or to relocate entirely.

Log report for Suwannee Band (Third Push)

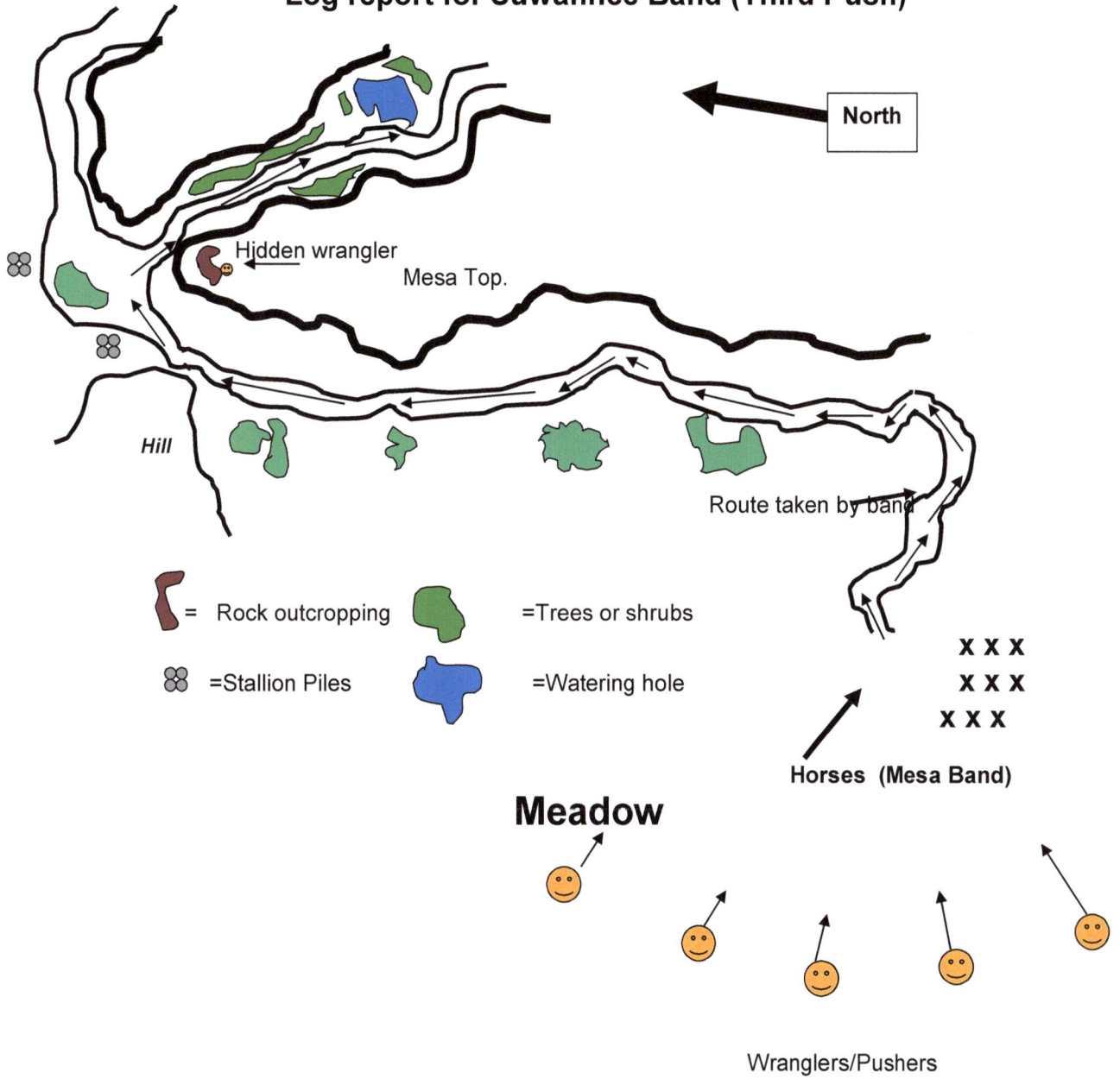

North

Hidden wrangler

Mesa Top.

Hill

Route taken by band

= Rock outcropping

=Trees or shrubs

=Stallion Piles

=Watering hole

X X X
X X X
X X X

Horses (Mesa Band)

Meadow

Wranglers/Pushers

Phase Two

The Round Up.

Having established the band's routes and escape trails and having determined which bands will be rounded up, the roundup must be planned.

First step:
Prepare your equipment.

The same equipment can be used for each band round up, but if you need to round up more than one band at a time, you will need two sets of equipment.

Equipment list

Portable Panels: X panels divided by **pen size.**
> i.e.,. If panels are 10' long, and the pen is going to be 50' x 100' you will need 30/10' panels.

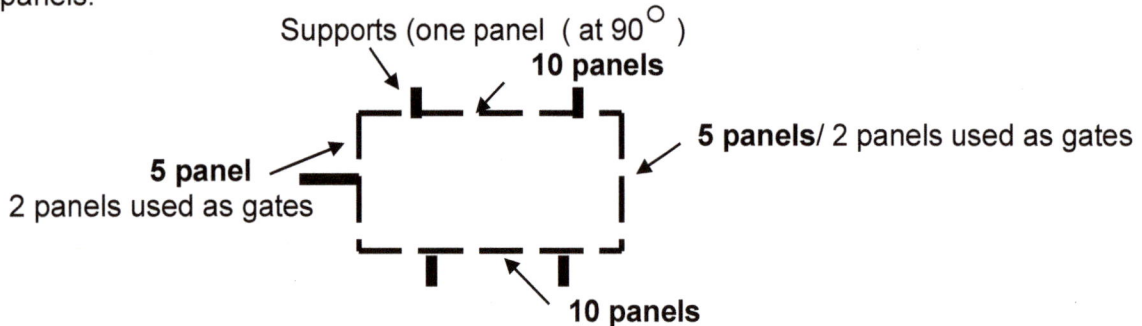

Supports (one panel (at 90°)

10 panels

5 panels/ 2 panels used as gates

5 panel
2 panels used as gates

10 panels

Determine if there will be wings on both ends of the catch pen. If so add another 40/10' panels to the equipment list. If only going to use one end wing then add 20/10' to the list. You will also need to place **support panels**, at least one for every four panels you set in a straight line. You should place the supports at the weakest points of the straight line from the corners.

i.e.:

On the wings, the support panels should be placed about one for every five straight panels.

.e.:

Note: It has been suggested that we might want to try Barrier plastic, the kind used by most highway departments when people do roadwork. We have not tried this so I can't comment.

Second step

Setting up the equipment.

It is important to know how to set up and tear down the catch pens and the wings.
In the field, this must be done quickly and without problems.

Remember: Noise is not a problem but loud startling sounds will cause the bands to become alarmed.

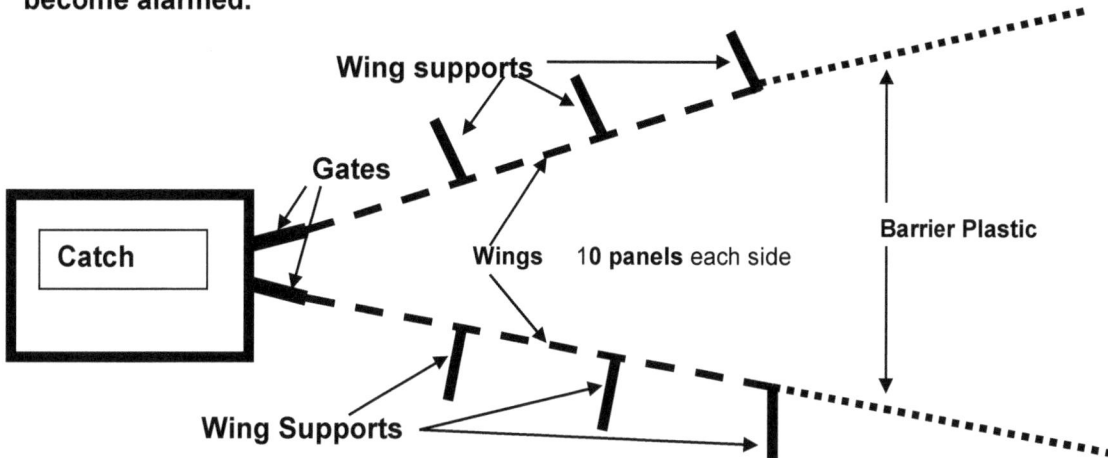

If Barrier plastic is used it should be of a neutral color with at least 100' long extension on each wing. As has been mentioned, we have not used this material but one of our wranglers, Randy Connors, believes that it will work; therefore, we will try it on our next round up. There is no set method to doing a roundup, just basic principles of safety, first yours then the horses. Consider all suggestions carefully; then see what might work. Do not disregard suggestions because they are new, remember "**fresh eyes**" see things differently.

Third step:

Getting equipment and yourself to the site.

This should be a given, but you would be surprised how easily this little matter can destroy a roundup. Imagine having 40 10' panels strewn across a freeway or down an arroyo because they are secured improperly. Check each vehicle and trailer's tires, gas, and oil. The last thing you need to do is run out of gas 50 miles out in the wilderness. This is the event you have been looking forward to for the past two or three weeks. You know the horses' movements, and you now feel you have a connection with each horse. Your excitement will be like that of a child on Christmas Eve, and some individuals have told me they could not sleep the night before the roundup. This is typical, so you must take extra steps to make sure you have not forgotten anything. There was one time an individual who was totally prepared got to the roundup site, worked all morning on the roundup, came in to get something to eat and realized he had forgotten to put his false teeth in when he left the house that morning. A hundred miles and five hours later he was back, but he did miss the first round up.

Fourth Step:

Setting up the equipment.

When you move the vehicles, don't worry about the noise you are creating. The horses by now are used to seeing and hearing you and the vehicles and probably will continue grazing. To be safe, though, take the longest route that maintains the farthest distance away from the grazing band you plan to round up. In some instances, depending on where the spot for the catch pen is located, you may have to carry the equipment a long distance. We have had cases in which we carried the panels for a mile and a half, which meant set time alone took two days. Under the circumstances, you might want to consider using saddle horses to help, but do not use them for the roundup.

Given that the above is not the case you now are ready to set up the catch pen.
Find the narrowest place on the trail you have determined to use for the roundup. Build your catch pen using a rectangular shape with at least 3 to 5 panels at the front or where the band will first enter the pen. From this point set up the panels on each side, in a straight line or as near to straight as possible, on a 90-degree line from the corner of the front panels. If it is possible, hide the back of the pen, if not don't worry about it.
Now that the catch pen is set up you must set up the wings.
When you set up the wings allow room for the gates to open and close. When you set up the wings you must leave room for the gates to the catch pen.

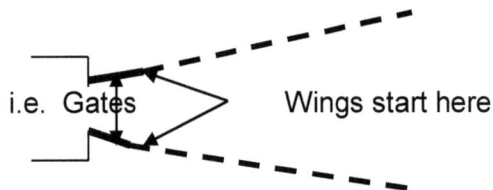

i.e. Gates   Wings start here

You must also be sure to have cover near the gates so that a wrangler can shut the gates after the band is in the pen. If there is no natural cover, you will have to establish a blind. If you have to establish a blind, make sure it is in place before the beginning of the third push. The band should be aware of anything new along the trail before the roundup. If you add new things along the trail, this could cause the band to turn back on the pushers and cause someone to get hurt.
Once the catch pen and the wings are set up, I think it is best to open the back of the pen. However, have a wrangler in position ready to close the panels if the band enters the wings. Then push the band one last time, make this push a slow, gentle push as on the second push. If the band enters the wings without any problems, the wrangler should close the back end of the pen. If the band balks at entering the wings, then stop the push. Leave the area and have lunch or get something to drink, but most of all, do not do anything to upset the horses. Give them time to get to feel comfortable with the new items on their trail. Patience is the key here; if you push them too hard, you may lose them for days. If you are patient, you will be able to round up the band within hours.
More than likely, you will be able to get the band into the catch pen on the first try, but be ready for a balk. There are many reasons for a round up to fail, but proper preparation eliminates most of those problems that we can control.
The result you are looking for is a band of horses in a catch pen without anyone, your crew, or the horses being hurt or stressed.

Log report for Suwannee Band (Third Push)

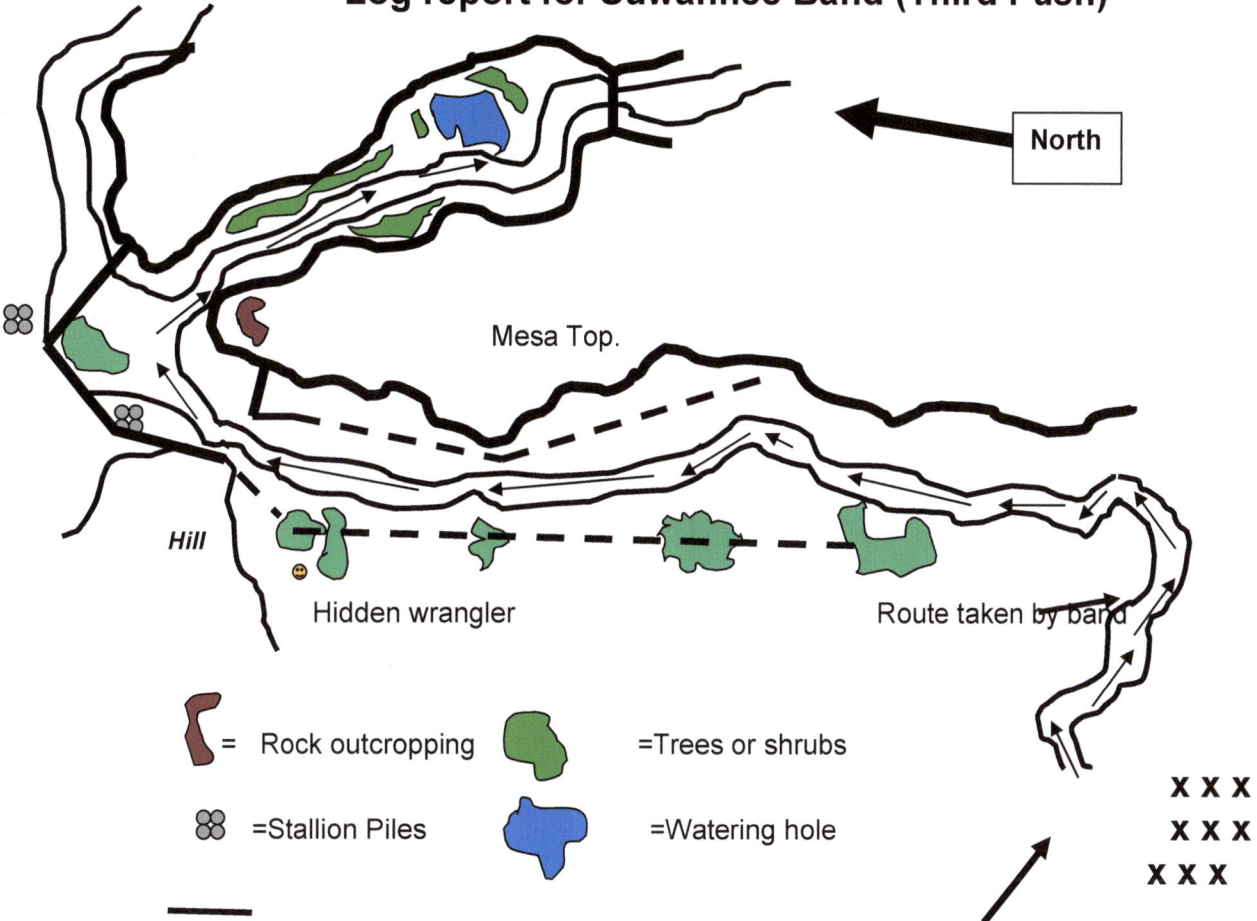

North

Mesa Top.

Hill

Hidden wrangler

Route taken by band

= Rock outcropping
=Trees or shrubs

=Stallion Piles
=Watering hole

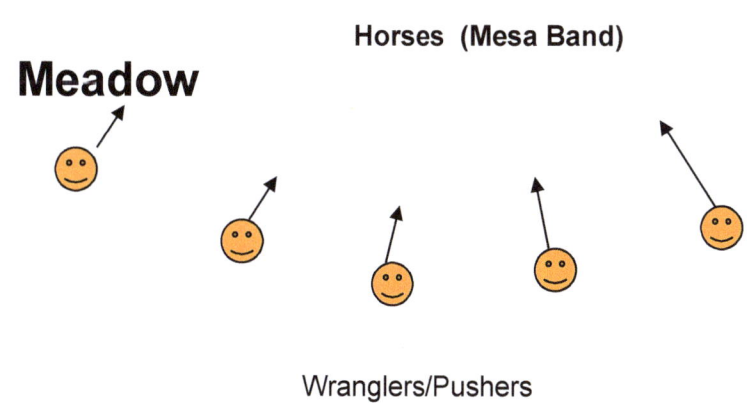

X X X
X X X
X X X

Horses (Mesa Band)

Meadow

Wranglers/Pushers

Log report for Suwannee Band (Third Push)

Blow up of catch pen

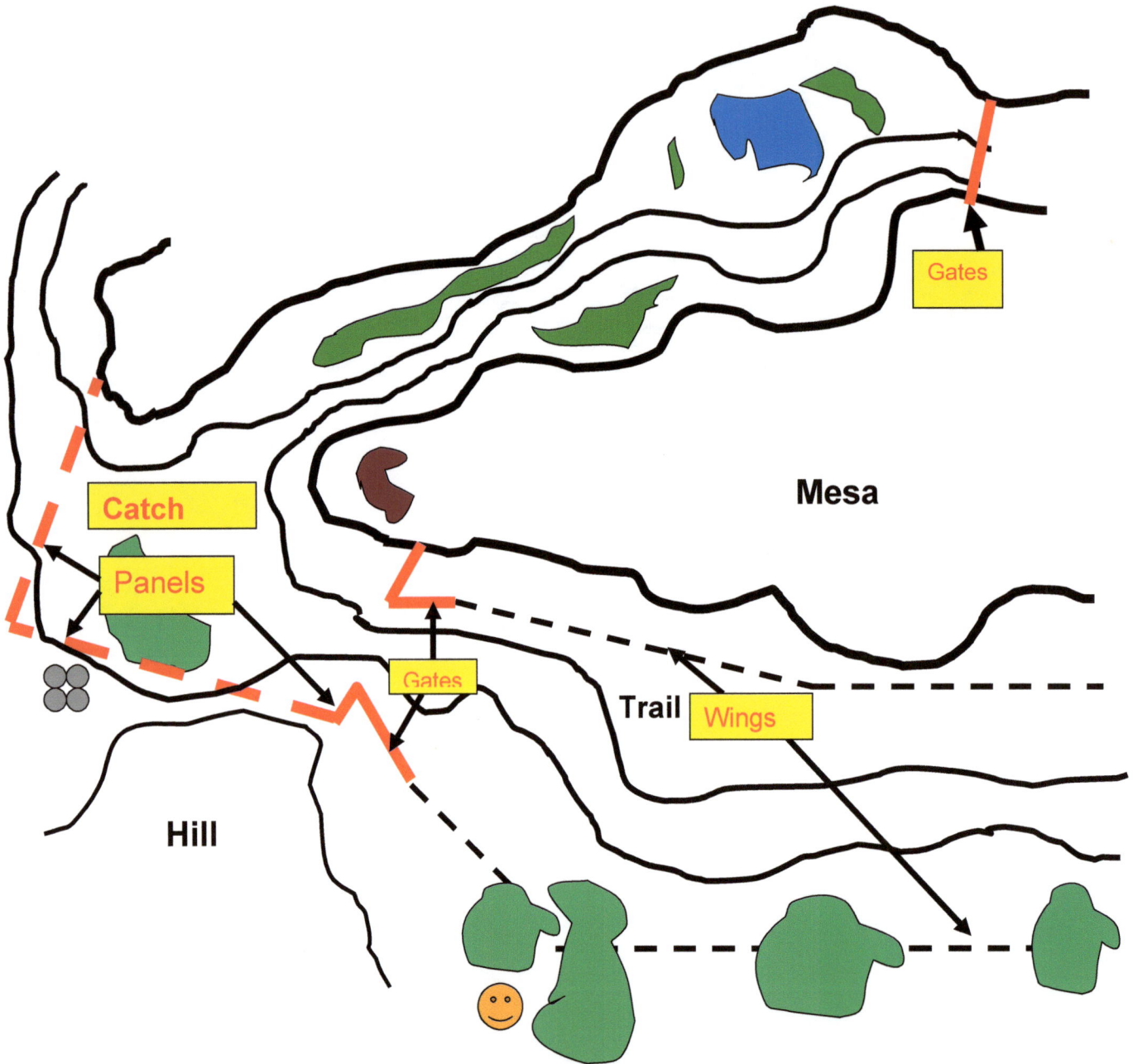

Gates

Catch

Panels

Mesa

Gates

Trail

Wings

Hill

Problems that could be encountered during the fourth step.

Problem: Because of all the noise and commotion caused during the set up of the catch pen and wings the band uses another trail.

Solution: Move everyone back to the staging area and wait for the band to come back. If you are lucky, they will return using the trail where the pen and wings are set up. This is why you leave the back of the pen open.

Problem: The band moves into the wings but balks when it sees the pen or something new.

Solution: There are two things you can do in this situation; one is to stop and wait for the band to calm down then continue the push. The other is to collapse the wings. If you collapse the wings you simply have created a very large pen.

Make sure if you do this you let the band relax and give them time to get used to their surroundings

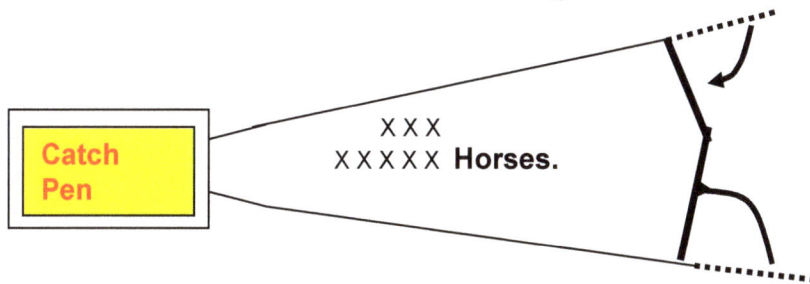

Catch Pen   x x x
x x x x x **Horses.**

Problem: The band takes another trail.

Solution: You can place wranglers along the other trails at the entrances to turn the band back. However, this is a two-edged sword because what your wranglers may do, through no fault of theirs, is to cause the band to scatter and move onto another bands' territory. Of course, you may be able to turn the band back to the original trail, and everything works out fine.

Step Five:

Celebrate. You know you have succeeded. You just did something that most people in the world only dream about. Of course, keep the yelling and "wahoos!" to a low roar, and try not to do it next to the corralled horses. The celebration should take 30 minutes to an hour. After you have calmed down and thought of the best way to tell your friends what you have just experienced, which is impossible by the way, it's time to get back to work.

The first thing the wranglers should do is walk around the pen and watch the reaction of the horses. See what they do when the wranglers get close. Check the condition of the band. Now that you can see them up close confirm what you thought about each horse.

Make sure you watch their eyes, be concerned if you can see the whites of their eyes, this shows they are fearful and can lead to them panicking. If you see this, back off and let the band calm down. This may take some time, so what you need to think about is making sure the band has water and feed.

Even if you only are looking at a few hours, it is best to have food and water available for the horses. This will also help to calm them. Sometimes we place the water and the hay in the catch pen before we start the push. It depends on the band and the opinion of the wranglers and is an "on the scene" decision.

You should have a 100-gallon water tank available and at least three bales of hay for each band you plan to round up. The same water tank can be used at each round up site, but the hay cannot. Make sure you use only grass hay, **<u>no alfalfa</u>**. **Alfalfa** hay can cause them to colic, as these horses are not used to this type of feed.

 After you feel the band has settled down and everything is calm, have one or two wranglers get into the pen and just sit down. Take a stool or chair and a book and just read. Watch what happens; you will be surprised. Within fifteen to thirty minutes, the horses will be exploring the wranglers. To me, this is the most exciting part of the roundup. The first real contact and probably the most important event to the rest of the operation. Make sure the wranglers who get in the pen have done this before because this could get dangerous and is not for the inexperienced.

- The first thing to happen is the band will huddle up in an area away from the wranglers, but will always keep them in sight. The wranglers at this point must not do anything, just sit and read or talk.
- It will take some time, but the band will start to move closer. If there are any young horses in the band, they will probably be the first to move near the wranglers, if not it will probably be the lead mare.
- When the band finally moves close enough to sniff the wranglers, which they will do, be careful and watch they do not bite. This is a natural thing for them to do. If one of the horses tries to bite, react the same way another horse would: confront or move. Whatever you do will cause the bitter to react and most likely stop.
- Once the horses are milling around the wranglers and have accepted their presence, have them leave the pen. It is important for the horses to know that you are not there to harm them. This will also help when it comes time to load them into a trailer.
- If the band has more than one stallion, separate them as quickly as possible. You will find that the out stallion might have a mare that stays with him; keep them together.

Blooding the Horses

Now that we have the horses in pens and have spent a day or two showing them we do not intend to harm them; we now must stick them with a needle. The intent here is to draw enough blood, usually from the jugular vein, which is located in the neck.
It is very hard to set up in the field and blood.   It is best to transport the horses to a secure pen that has at least one solid wall.

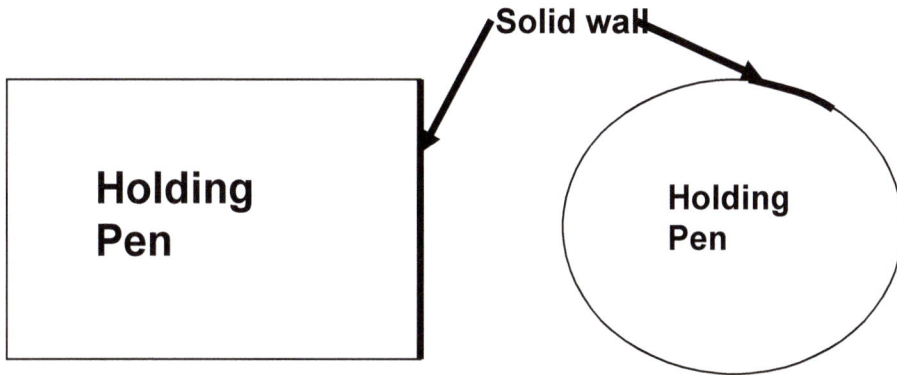

Solid wall

Holding Pen

Holding Pen

It does not make a difference what the shape of the pen is as long as it has one solid wall.

Set up a turnout pen next to the holding pen for the horses to go into when blooding is completed. As you complete the blooding of one horse, turn it into the turnout pen.

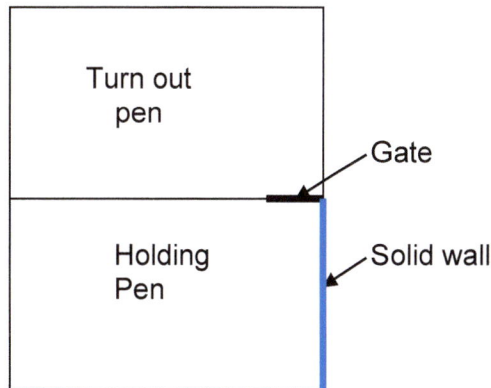

Turn out pen

Gate

Holding Pen

Solid wall

After you have the horses in a secure pen and the turnout pen, you must set up a squeeze chute.
This is done by taking one panel and forming a wedge using a solid, secure panel or wall.
A secure panel must anchor the wedge and be rigid and sturdy; if this panel fails because of movement, any movement, the horse will feel it can escape and will never calm down. If this happens, the technician will not be able to draw blood.

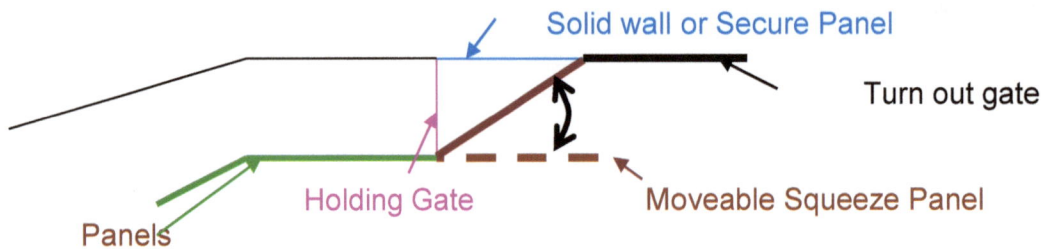

Solid wall or Secure Panel

Turn out gate

Holding Gate

Moveable Squeeze Panel

Panels

The panel that will get the most work will be the squeeze panel, and it must be able to open and close as seen in the above figure.

It does not make any difference which horse you blood first. However, always try to keep a mother and her foal together. If not, it will cause the mother stress and will alarm the rest of the horses.

The horses will enter the pens the way they want to, just work with the horse as it comes in. Once the horse has entered the squeeze pen, if you have a gate at the end of the squeeze panel close it so the horse cannot back out. Sometimes we do not have the luxury of a gate, so we use a pole behind the horse's rump to stop it from backing up.

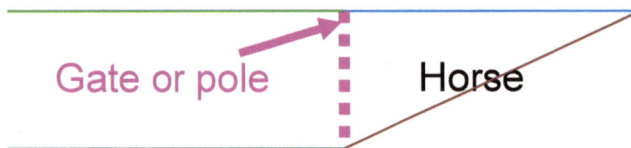

Gate or pole

Horse

If the pole is under the tail set, it will usually stop any backward movement. If it is too high, the horse will back under it and too low he will walk over it. It must be cautioned now that the horse is in the squeeze pen, the technician is at the most dangerous place in the roundup. The technician will be standing next to a wild horse with his arms extended into a pen trying to find the horse's jugular vein and draw blood. Do not do anything that will create a stressful situation. It is best to move away from the area and his or her helper, stay quiet. Once the blood is drawn, place it in the prescribed vials that come with the testing kits and mark it appropriately. It is always best to take a picture of each horse during the blood draw. After the blood is drawn and placed into the vials, the vials must be coded.

Coding the Vials

During the first phase when you were spotting the bands you probably named each one. The Pinon Hill band, the Paint band, etc. Use that name in the code.
For example, PHB-1 would mean the first horse blooded in the Pinon Hills band.
You also should describe the horse. Give every marking you can to show the uniqueness of the animal.

We use this form.
You may want to develop your own form to suit your needs

Date of Capture_____ Date of Blood Sample_____

Vet's Name_____ Recorder's Name_____

New Mexico Horse Project #_____ NK#_____

Sex_____ Relative Age_____

Location_____ Herd_____ _____ Band_____

Coloration: Body_____ Head_____

Eyes_____ Mane_____ Tail____ _____ Hooves_____

Stockings_____ Comments_____

Scars_____ _____

Health_____ _____

SHOW ALL SIGNIFICANT MARKINGS, WHORLS, BRANDS, AND SCARS

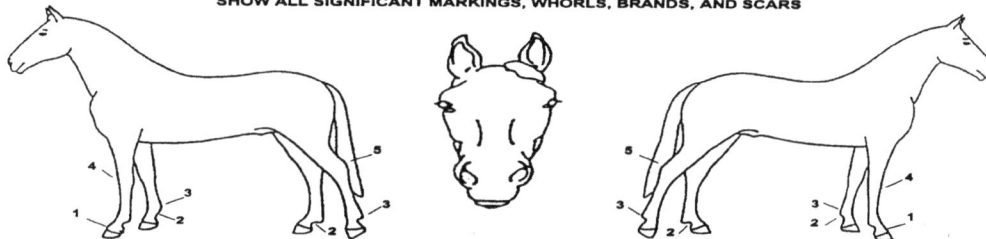

1 - Coronet, 2 - Pastern, 3 - Fetlock, 4 - Knee, 5 - Hock

NARRATIVE DESCRIPTION AND REMARKS

LEFT FORELIMB	RIGHT FORELIMB
LEFT HINDLIMB	RIGHT HINDLIMB

Make sure you mark each picture with the code, or you may lose your information or get it mixed up, and your blood draw process will have been for naught. It is best if you can video the process also.

**Turn out pen
for after blood
draw**

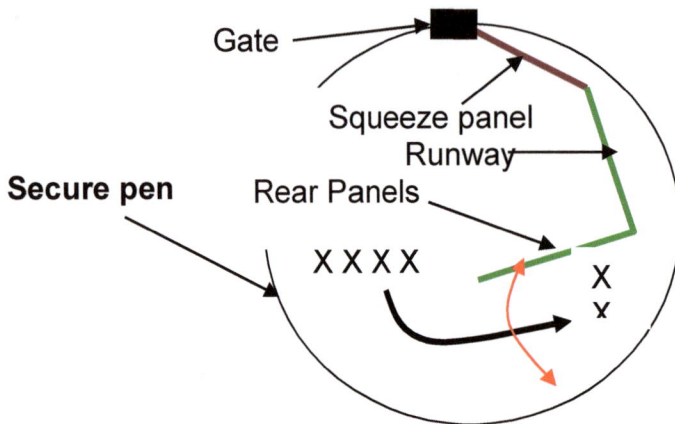

Gate

Squeeze panel

Runway

Secure pen

Rear Panels

X X X X

X
X

Once the horses are in the runway close the
rear panels and
form a small pen within the larger pen.

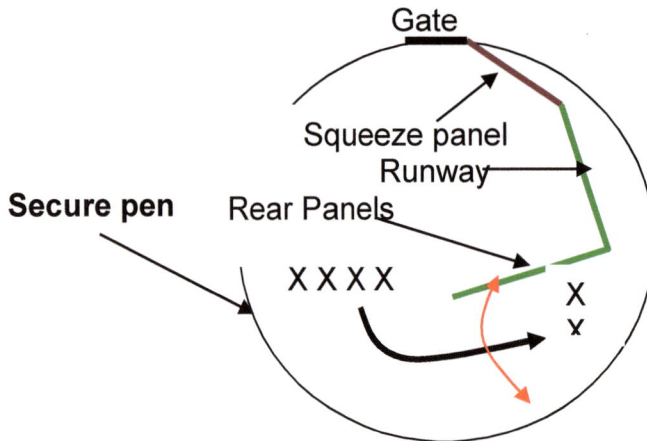

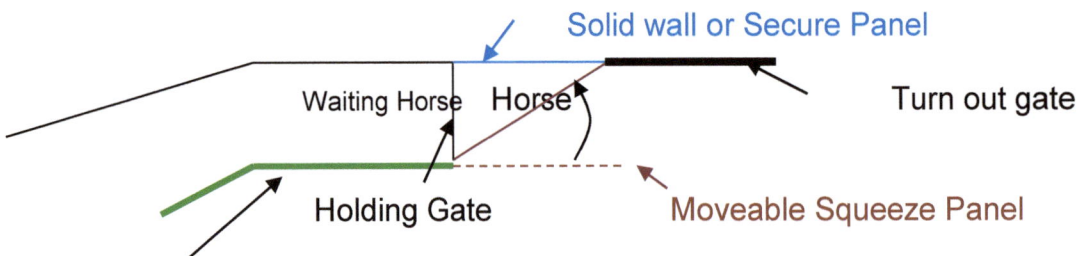

Gate

Squeeze panel

Runway

Secure pen

Rear Panels

X X X X

X
X

Once the horses are in the runway close the
rear panels and
form a small pen within the larger pen.

Solid wall or Secure Panel

Waiting Horse   Horse   Turn out gate

Holding Gate   Moveable Squeeze Panel

TERMINOLOGY AND DEFINITIONS

Band: A group of equis usually not larger than 10. Consisting of a lead stallion and his mares and their young. The number 10 is not a hard, fast rule but only an average. Bands sometimes are as large as 15 and as small as 3. It depends on the stallion and how many mares he can protect or how dominant he is.

Band maps: Maps developed to depict the routes/trails used by the horses when pushed and showing each bands territory and boundaries.

Blind: Man-made hiding place.

Bleeder/Technician: One who draws the blood.

Blooding: Drawing blood for the purpose of DNA testing.

Catch Pen: The pen used to capture wild horses. Usually made of portable panels but can be made of wood or other materials strong enough to hold the horses.

Colt: A male equis that has not reached the age of being capable of reproduction.

Cover: A place to hide from the horses.

Escape route or trail: The route or trail used by the band to escape intruders.

Filly: A female equis that has not had an offspring.

Fresh Eyes: Someone who is seeing things for the first time.

Gators[1]: Protection that wraps around ones' ankles and calf muscles to protect the individual from snake bites

Gelding: A male equis that has been altered so that it cannot reproduce.

Hand: a unit of measurement being four (4) inches in height.

Herd: A group of bands of equis that is located in a specific area and has been confined to the area because of natural or man-made barriers.

Horse: Now means any equis, During the Spanish Colonial period this term was used to represent a stallion

---

[1] Gators is a trade name of these protectors.

Intruder band: A band of horses that travel into another bands territory.

Lead Band: The most dominant band in a herd consisting of the most dominant stallion of the herd and the most dominant mare of the herd. This band gets the first choice of the best grazing and the best watering holes. The equine in this band appear to be in the best shape and the healthiest.

Lead Mare: A female equis that is the strongest or most controlling female of the group.

Line of Demarcation: An imaginary line established by the horses that when crossed by a pursuer the horses will move away.

Mare: A female equis.

Natural Cover: A place to hide that is part of the environment and not man made.

Portable Panels: Barriers that are usually made of metal used to confine horses or cows. They can be assembled in the field and are not fixed to one place.

Push or Pushing: Moving the horses from one place to another.

Stallion: A male equis that can reproduce offspring.

Stallion Piles: A pile of manure placed in a specific area by the lead stallion of a band to mark the boundaries of the territory he controls.

Support panels: Panels that are used to support a straight line of panels. Support panels are set at a 90-degree angle from the straight-line panels.

Walking them Down: Rounding up horses on foot.

Wind Strap: A strap that is put on a hat that goes around ones' chin and prevents the hat from being blown off.

Wings: Panels that protrude from both sides of the entrance of the catch pen to help guide the horses into the pen. Usually 100 feet to 200 hundred feet long.

Wild Horse: A horse that has had little or no contact with humans during its life.

Wrangler: Anyone working on the roundup that has had training.

Wrangler Training: [2] This means being taught how to round up horses on foot or by "walking them down." Anybody can run a horse using a 4x4 or another horse or even a helicopter, but a real "**wrangler**" is one who can "**Walk them Down.**"

Yearling Colt: A male equis under the age of one year.

Yearling Filly: A female equis that is under one year of age.

---

[2] No one should be allowed to take part on a round up without proper training. For the New Mexican Horse Project this meant knowing how to round up horses on foot or "walking them down."

www.ingramcontent.com/pod-product-compliance
Lightning Source LLC
Chambersburg PA
CBHW060808270326
41927CB00003B/89